I0470660

This report is part of a series of reports on technical rescue incidents across the United States. Technical rescue has become increasingly recognized as an important element in integrated emergency response. Technical rescue generally includes the following rescue disciplines: confined space rescue, rope rescue, trench/collapse rescue, ice/water rescue, and agricultural and industrial rescue. The intent of these reports is to share information about recent technical rescue incidents with rescuers across the country. The investigation reports, such as this one, provide' detailed information about the magnitude and nature of the incidents themselves; how the response to the incidents was carried out and managed; the impact of these incidents on emergency responders and the emergency response systems in the community; and the lessons learned. The U.S. Fire Administration greatly appreciates the cooperation and information it has received from the fire service, county and state officials, and other emergency responders while preparing these reports.

This report was produced under contract EMW-94-C-4436. Any opinions, findings, conclusions, or recommendations expressed in this publication do not necessarily reflect the views of the U.S. Fire Administration or the Federal Emergency Management Agency.

Additional copies of this report can be ordered from the U.S. Fire Administration, 16825 South Seton Avenue, Emmitsburg, MD 21727.

Search and Rescue Operations in California
During Flooding
January to March, 1995

Local Contacts:

Glenn Gibbons,
 Rescue Officer
Camp Pendleton Fire Department
Camp Pendleton, CA 92055
(619) 725-4111

Mark Ghalarducci,
 Division Chief
California State Office of Emergency Services
2800 Meadowview Road
Sacramento, CA 95832
(916) 262-1685

OVERVIEW

During the first three weeks of 1995, California experienced the second highest rainfall of the 20th Century. Emergency officials prepared to respond to the flooding that would easily equal and surpass similar periods during the winters of 1992 and 1993.

During 1995, however, the only similarities throughout the state were the overflowing flood control system and the inundation of large numbers of houses in Napa, Sonoma, Sacramento, Placer, Orange, Los Angeles, and San Diego Counties.

In contrast, local and state emergency agencies had spent the previous two-year period preparing to respond to just such events. Thousands of rescuers had received at least first responder level training for dealing with moving water rescues. Equipment had been purchased. In many areas hit in 1992, particularly in Los Angeles County and its environs, major pre-planning had been undertaken. Coordinators knew exactly what resources were immediately available. Task forces had been designated, had trained together, and were activated as soon as rainfall pointed to the chance of localized flooding.

The result was a substantial decrease in the loss of life compared to similar periods in 1992 and 1993. There were six deaths during the flood of January 1995 in California - contrasted with comparable rainfall in Georgia during a like period (about eight days in July of 1994) that lead to 35 deaths. In the March flooding that lasted nearly two weeks, there were 11 storm-related fatalities. Of these, two were skiers who died in a blizzard near Lake Tahoe and therefore not specifically flood related.

The rainfall in California was due to a series of huge storms originating in the Pacific Ocean, and exasperated by the weather condition know as "El Nino." (This is a mysterious and periodic effect that warms the Pacific Ocean near the Equator, bringing sustained wet weather to Mexico and the entire southwestern United States.) In January 1995, the series of heavy storms lasted nearly three weeks. In March, storms marched ashore every three days for nearly two weeks. In the intervening few weeks of dry weather the ground remained heavily saturated, compounding problems as structures, houses, bridges and entire hillsides finally gave in to nature.

The United States has experienced several major floods in recent years. Flooding along the Mississippi River in 1992 affected a substantially larger area and caused more severe property loss. The Georgia flooding was of shorter duration and caused more loss of life. North Carolina flooding in August 1995 also caused major local disruptions.

The devastation caused by the California flooding on the economy, property loss and loss of life falls somewhere in the middle of the above events. Still, the results of such sustained rains were impressive:

- Thirty-eight counties declared states of emergency. Ten remained under continuous states of emergency from the first week of January to the end of March.

- Even in the 13 counties that did not declare emergencies, there were still storm-related problems, including flooding, mud slides, heavy snows, local road closures, and river rescues.

- In Napa County, nearly 300 homes were flooded, many of which were in the process of being repaired from the January flooding. Several thousand people were evacuated from their homes there as well as along stretches of the Russian River in neighboring Sonoma County.

- Sodden soil combined with continuous flooding and washed away a bridge over Interstate 5 near Colinga in Fresno County, three cars went into the usually dry creek bed. One victim was saved, but three others died.

- Residents in Carmel Valley had to be evacuated in Monterey County as the Carmel River went over its banks, and took out the only bridge across Highway 1. In the northern part of the county, the entire population of the little town of Pajaro and many residents of Castroville had to be evacuated.

- Highway 17 in Santa Cruz County, 200 miles of I-5 in Kern County, and stretches of Highway 101 in Los Angeles County were closed. Cumulatively, road and highway closures effectively paralyzed the state for several days, leading to incalculable numbers of closed businesses and lost commercial dollars.

- Total damage state-wide was estimated at $3 to $5 billion.

I. EMERGENCY RESPONSE

Overall, January's three weeks of rain impacted Southern California more substantially than Northern California. In March just the reverse occurred, with Northern California receiving more direct impact over a ten-day period starting in early March. Therefore, to examine state-wide emergency response over a three-month period, it is best to look at local responses, starting in Southern California and moving north. Conclusions can then be drawn from similar incidents and responses.

San Diego County

Prior to 1992, the San Diego Lifeguard River Rescue Team was the lead agency for flood response. The Rescue Team dealt with nearly 200 incidents in California and Mexico during the 1992 floods.

Since 1992, other agencies in the area have instituted training programs and a county-wide training group has been formed by interested agencies. The results started to pay off in January 1995.

Lifeguards saved 21 victims, including nine saved on a night operation in a narrow canyon. Most of the rescues involved vehicles driving around barricades and becoming stuck in the middle of low-water crossings. One female victim perished in the flooded basement of her home, while rescuers attempted to gain access from the floor above.

In March near Oceanside, swiftwater trained firefighters from nearby Camp Pendleton Fire Department joined with base lifeguards to search for a Marine Corps colonel who was washed away as he tried to lead his troops across a flooded stream. His body was found the next day nearly six miles downstream, pinned against debris.

Orange County

The 1993 brush fires blackened large coastal areas, including homes and acreage around Laguna Beach. In January, as rainfall approached an inch each hour, the Orange County Emergency Operations Center alerted the swiftwater rescue trained joint-agency teams to evacuate downtown businesses and a preschool in Laguna Canyon. Lifeguards again played a key role, making themselves available to be the strong swimmers if necessary.

Los Angeles County

Los Angeles County agencies experienced most of their major problems in January, although several rescues were made during the March flooding. Riverside and San Bernardino Counties experienced flash flooding as well, particularly in the their eastern areas.

Prior to 1992, individuals from numerous agencies in the Los Angeles basin had taken Swiftwater Rescue Technician (TM) training, or had participated in in-house programs on river rescue.

However, the 1992 floods revealed the lack of area-wide preparation. Vast areas of the Sepulvada Basin suddenly went underwater, trapping large numbers of motorists in their cars. In another horrifying incident, a 15-year-old boy, Adam Bischoff, drowned after managing to stay afloat while being swept down nearly 10 miles of the flood control system, which contains more than 400 miles of channels.

As a result, a joint agency task force was founded not only to meet training needs, but to establish standards for equipment and pre-plan for future problems. The task force consists of law enforcement, fire, and lifeguard agencies from the city and county.

After instituting training and national certification for instructors, those same instructors trained more than 3,000 personnel as technicians or first responders.

Los Angeles Fire Department swiftwater instructors Ernesto Ojeda and Jim Goldsworthy prepared detailed maps of the 470-mile flood channel system and prepared references for preferable rescue techniques at potential rescue sites. Working with analyst Kathy Evansen, they then created a computerized dispatch system to allow dispatchers to match rescue assets with location, depending on where a victim enters the particular flood channel and the amount of water flowing.

In January, and again in March 1995, Los Angeles County received more rain than the average yearly total! In both months, despite the huge amount of water, the Swiftwater Rescue Task Forces were pre-positioned, evacuating hundreds of people and making more than 40 swiftwater rescues. There were still five fatalities, including a car that went off the Pasadena Freeway into the Arroyo Seco channel, and a transient who died after being washed away in a flood channel in the Sepulvada Basin.

Ventura County

In 1992, Ventura County agencies had several rescues involving firefighters. One such incident occurred when a lieutenant became trapped in midstream. Another happened when a hovercraft flipped and washed away during flood operations along the Ventura River.

By 1995, both city and county departments had trained some members in swiftwater rescue. Ventura County Fire Department had sent one member to become a swiftwater rescue instructor. As the result, every city firefighter, a large number of county firefighters, and members of the Sheriffs aviation unit and rescue team had been trained to at least swiftwater first responder level.

As in 1992, the Ventura River again in 1995 was the hot spot with trailer homes needing to be evacuated. Even though they had been warned earlier to evacuate, homeless people along the river needed to be rescued. On January 10, 1995 Captain Myles Smith of the Ventura City Fire Department used his swiftwater training to make a swimming rescue of a victim who was trapped in a debris pile and in imminent danger of being pulled under and drowned. Putting him in a special river rescue jacket, Captain Smith swung the victim on a rope to shore near other rescuers.

Shortly afterwards the calls for rescues started in earnest. By the end of the day, combined air and ground units had rescued more than 35 people and half a dozen dogs and cats. There was only one human fatality.

Protracted operations for rescuers continued in the small bedroom community of La Conchita. As a hillside behind the homes become more unstable, residents were evacuated. Nine homes were destroyed by mud and sliding earth. Initial geology reports indicated that all of the remaining homes might have to be condemned, but homeowners were later allowed to return.

San Bernardino County

According to Fire Captain Andrew Bennett, during the March 1995 flooding three boys were swept down a flood channel for nearly 1 1/2 miles before rescuers could safely retrieve them. While flooding was widespread in January 1995, most operations involved evacuating flooded residents.

Santa Barbara County

The January floods had only a moderate impact on Santa Barbara County, but March rainfalls exceeded levels in recorded history. Santa Barbara City Fire Department swiftwater instructor Guy Smith's training efforts paid off as personnel responded to dozens of calls for assistance in Santa Barbara.

Despite their efforts, one man, a retired federal judge, died when he was swept from his Sycamore Canyon home. His wife survived by clinging to a light fixture in the home. Eighteen-year-old Rosa Jiminez disappeared after trying to cross a flooded creek in her car. Unlike drivers in two similar vehicle incidents, she tried to escape by swimming away from her trapped car.

In Lompoc, rescuers used inflatable rafts and ropes to rescue two persons from their vehicle when it was washed away in the Santa Ynez River.

San Luis Obispo County

While road flooding was widespread, the only significant rescue found firefighters trying to rescue their own chief after his department car was washed away in a low water crossing. The officer was successfully saved using an inflatable boat and ropes.

Monterey County

The Monterey, Salinas, and Pajaro Rivers rose again, as they had in February. In March the problems were much worse due to saturated ground.

In Carmel, the river flooded out the Highway 1 Bridge, leaving only a detoured route for residents south of the river to evacuate. Only a few trained river rescuers were available, and firefighters and sheriffs officers scrambled to evacuate residents in the Carmel Valley area. While occasional high water problems have occurred in the past, this was the worst flooding along the river in recorded history.

Santa Cruz County

Thirty-three hundred people were evacuated from Pajaro when the river rose seven feet over flood stage. Twenty-seven hundred people were evacuated from low-lying areas near Watsonville, which was still recovering from the Loma Prieta earthquake. A levee break fortunately alleviated the pressure to the San Lorenzo River.

In a repeat of 1992's problems, localized mud slides closed stretches of Highway 17 connecting Santa Cruz with San Jose. Boulder Creek also rose to record levels in the Scotts Valley area.

Despite numerous evacuations, there were few true rescue situations.

Santa Clara County

With Lexington Reservoir already full from January 1995, the March rainfall created numerous problems in San Jose, which experienced its most substantial flooding problems since 1987.

By March 11, the normally nearly dry Guadalupe River overflowed putting four feet of water into large areas of downtown.

While dealing with those problems, fire personnel and rescuers were dispatched to Alamitos Creek, along with a helicopter. They were to search for a missing teenager who had attempted to kayak in the floodwaters after receiving permission from his parents. After flipping and losing his kayak he had managed to rescue himself, but rescuers were not aware he was safe for several hours.

Alameda County

Fremont Fire Department sent a heavy rescue assignment to Niles Canyon on a report of a rafter stuck in a tree. Untrained and unequipped for river and flood rescue, firefighters initially called an East Bay Regional Parks helicopter for help. Unable to make a one-skid landing on the debris pile where the victim was stranded, rescuers finally had to send the helicopter away due to the noise level, and their own inability to hear their radios.

Firefighters found a wetsuit and two horse collar lifejackets left by the stranded victim's friends. Tying one firefighter to a rope, they finally managed to throw the victim a line and swing the victim to shore.

Unlike other areas with similar problems, fire personnel later said, (when interviewed for a fire video training series), that while they might buy some river rescue equipment, they doubted they would do any specific river training, due to the infrequency of such calls.

Fresno County

Initially unaffected, except for localized flooding by the January 1995 rains, Fresno County was the scene of at least two notable flood rescues in March.

Friday evening, March 10, at approximately 9:00 p.m. nearly 150 feet of Interstate 5 was washed away when a bridge spanning Arroyo Pasajero Creek failed, near the west valley's already saturated soil. The failure was initially attributed to heavy water flows combined with the liquefaction of already saturated soil.

Initially motorists were able to stop, but three cars ignored the frantic waving of bystanders and sailed off the edge at full speed. One driver, Wayne Johnson, recalled that his windows would not go down or the doors open, so he exited through the rear window of the cab of his pickup and ended up clinging to a tree in the frigid water for nearly an hour, before a volunteer rescuer helped him to shore.

The other drivers were not as fortunate, and the final toll was four dead. One of the cars was found 200 yards downstream by rescue team personnel, on its roof and packed with mud.

Besides the fatalities and rescue operations, this incident effectively eliminated one of the two remaining north-south routes in the state, closing 200 miles of freeway which normally carry more than 25,000 cars a day. After initial predictions that the freeway would be closed for several months, a contractor was able to build a temporary bridge out of railroad flatcars, and limited traffic was restored in less than two weeks.

A little over one week later on March 19, 1995, a 7-year-old from Clovis, near Fresno, died after the canoe she was riding in with her 4-year-old brother and two uncles hit a bridge abutment on the swollen San Joaquin River. When the boat flipped, the 17 and 24-year-old men, who were not wearing lifejackets, were quickly rendered unable to swim by the 44-degree water. Swiftwater-trained Fresno firefighters and the American Ambulance Special Trauma and Rescue (STAR) team could not lift the girl from the debris pile until one of their members climbed down and cut her free from sticks and debris hanging on to her lifejacket.

The California Highway Patrol (CHP) rescued the 17-year-old and 4-year-old about 1/2 mile further down the river with their helicopter.

Sonoma County

Because of its close proximity to the Russian River, Sonoma historically has suffered during floods, and January and March of 1995 were no exception.

Nearly 200 homes in Guerneville and Forestville were inundated during both periods. Many homeowners and businesses had just started re-conditioning their buildings when the March rains began.

Evacuations and rescues occupied firefighters as well as sheriffs rescue team members and sheriffs and CHP helicopters, as far south as Petaluma.

The only fatalities reported in the area were both drivers killed in their vehicles by water-eroded trees falling on them. Several vehicle drivers were rescued by personal watercraft and inflatable rescue boats, and others were pulled to safety by helicopters "hot landing" on the roofs of the trapped vehicles.

Napa County

January and March were equally disastrous for Napa County. In one instance rescuers found themselves needing to be rescued when their boat washed away while trying to rescue a woman trapped in waters raging through a vineyard.

Elsewhere in Napa, 200 homes were flooded and nearly 200 deluxe mobile homes went under water in Yountville. Initial losses were projected to exceed $36 million.

County dispatchers reported that call volume increased by 300 percent when Napa Creek overflowed into downtown Napa. With more than 40 calls backed up for rescue services, radio frequencies were so overwhelmed that dispatchers had difficulty giving out the call.

Several children were rescued after attempting to raft the river - several others were stopped before launching and their rafts were confiscated. As in Sacramento County, local rescuers made good use of inflatable rescue boats and a dump truck. "Dump 1" was dispatched to motorists stranded in intersections in several parts of the city.

Learning from the floods of 1986 and 1987, officials evacuated more than 1,200 residents in various parts of the county to pre-positioned shelters. Three Emergency Operations Centers were established to distribute the call load - unfortunately, they had trouble communicating with each other, due to the overload of the phone and radio systems. Dispatcher Julie Hubert pointed out the value of using the media to keep people informed and to pass on evacuation information, while reminding people not to call 911 except in cases of true emergencies.

Sacramento County and Placer County

Northern areas of Sacramento County and western Placer County suffered the worst of the January rains. Normally placid Dry Creek in Roseville, rose suddenly and inundated several hundred homes along its normally 10-foot-wide channel.

Flood rescues occurred throughout early January, prompting swiftwater teams from the area fire departments and the state Office of Emergency Services (OES) Diving Accident Rescue Team to respond with boats, inflatables and personal watercraft to rescues for people stranded in cars, trees, and houses.

As the rains continued and calls for service increased, the state OES started requesting mutual aid resources from neighboring counties for not only the Sacramento area, but Napa and Sonoma Counties as well.

In some instances the most appropriate assets were not utilized. In one river rescue a sheriffs patrol boat lost its propeller in the shallow waters, and washed away. Fortunately, the officers made it to shore downstream.

In another situation, officials sent an Army Engineers unit to help evacuate residents of a retirement home surrounded by water. The 30-foot landing craft, however, had nearly twice as deep a draft as the water in the parking lot. Residents decided to wait where they were and let the water recede.

Residents of the areas hit hardest in January were ready in March. They started sandbagging and evacuating early, so problems were minimal.

However, in the March rains, neighboring communities along Cache Creek, the Sacramento River, and spreading into neighboring Yolo and Colusa Counties, were inundated by rising water. The Sacramento County Sheriff's helicopter unit found itself running an evacuation service from a small "island" in Rio Linda to nearby evacuation areas along I-5.

As in January, fire and rescue agencies found units being dispatched frequently to rescue children and rafts in local ditches. Many of these reports were determined to be false, and no lives were reported lost.

II. LESSONS LEARNED

The following lessons learned are from the incident commander's perspective:

1.   *Recognize the potential natural hazard and pre-plan accordingly.*

    Flooding is an eventuality in virtually every populated area in the United States. Budgeting for training and equipment should not be determined by what happened most recently, but rather by what is eventually going to happen.

    Ten years of drought found California agencies unprepared to respond to flood rescues adequately in 1992. Long oriented towards preparing to respond to earthquakes, officials had to respond to criticisms towards their operations during those floods. As a result, many agencies took a more multi-task perspective with their rescue units. In 1995 the state was well-prepared with trained rescuers and equipment. The next flood will find the state even better trained and equipped.

    The result has been that recent floods have still caused substantial property loss, but less loss of life.

    Be Pro-Active, not Reactive.

2.   *Mitigate the hazard.*

    Local planners and officials can diminish flood and river rescue problems by making regulatory decisions - some easy to take, and some hard:

Public education on the dangers of flooding, starting at grade school level can alter public behavior. Drivers will learn why they must avoid low-water crossings. Residents will understand why they <u>must</u> evacuate when ordered. And children can be taught why they must stay out of flooded rivers and flood channels. "NO Way Out" is one example, a video produced by Los Angeles resident Nancy Rigg, who witnessed her fiance's death attempting a river rescue. "No Way Out" is presented to school children throughout the Los Angeles area. It was designed to focus the public's attention on flooding dangers thereby minimizing such events from occurring in 1995.

In addition, reinforce the message negatively. Drivers moving around barricades should be heavily cited. Refusal to evacuate should be grounds for citation as well. Law enforcement officials and judicial officials should stand by such citations. They cost money, but save lives.

Local building and community planning can diminish rescue problems and save billions in tax dollars. Flood channel systems in California have been overwhelmed in recent rains. The use of federal tax dollars to repeatedly rebuild homes along flood-prone rivers has generated public criticism.

Some hard but effective decisions include: control growth in flood prone areas, institute tougher building codes on rise of foundations, and require insurance in order to secure building permits.

3.   *Identify and secure training resources.*

There are several credible training companies, groups and programs in the United States able to provide river and flood rescue training. Minimum consensus standards on training exist. Such training will not only save lives, but identify potential rescue problems, and preserve rescuer safety.

4. *Rescuer safety is the first priority.*

Between early 1983 and September 1995, 37 firefighters and rescuers have drowned, an average of three a year. Many of these drownings were preventable with some knowledge of the forces at work in moving water. Again, many of these deaths were due to a reaction to a problem and a well-intentioned, but misinformed, effort to save a life.

In river rescues the hard decision for the trained rescuer is to sometimes simply stand and watch, rather than endanger a rescuer. Again this decision is based on training and knowledge.

5. *When responding to flood rescue calls, send a full complement of resources immediately.*

Moving water rescue calls require as much skilled rescue power as possible, and the situation may change quickly and radically. Many failed rescues have been the result of sending only a "first-in" unit to evaluate the need for further help. By the time adequate help arrives the victim has washed away.

6. *Establish and utilize state and local mutual aid lists, and use them.*

A disaster is a situation that overwhelms local ability to respond. California's response to the floods of 1995 shows the value in working with neighboring agencies to establish inter-agency rescue task forces. It also showed the value of having state resources available.

7. *Emergency managers at the state level should understand river and flood rescue needs, and maintain a state resource list.*

State emergency planners can support tactical operations by maintaining lists of trained river rescue units, boats appropriate to flood rescue, and helicopter units trained in moving water rescue work.

Sending inappropriately trained resources to a scene may slow the rescue response, as well as possibly endanger rescuers inadequately trained for the environment.

8. *The federal government can provide for recovery efflorts, but river and flood rescues will be done by local agencies.*

Even state level resources take time to respond. By the time outside resources arrive, the majority of the rescue work will be over, good or bad.

9. *Use the correct equipment for the job.*

Coast Guard approved personal flotation devices, either types III or V, are an absolute minimal need for any rescue personnel who are near the water. Personnel floatation devices should not be allowed to approach the water. Further, river rescue experts generally concur that fire "turn-outs" and bunker boots are potentially dangerous if a firefighter is washed away.

10.   *Incident commandplanning documents and flip cards need to identify swiftwater and flood assignments.*

Incident Command System (ICS) card systems generally tell incident commanders what to request for various kinds of calls, including earthquakes, tornadoes, wildland fires, etc. These card systems need to include flood and river responses as well.

11.   *Establish inter-disciplinary and inter-agency task forces.*

In California, local areas in both north and south have gotten past political concerns in order to establish task force plans. The task forces utilize appropriate resources from police, fire, sheriffs, lifeguards, ambulance, dive teams and rescue teams. Incident commanders need to know the capabilities of each of these groups.

12.   *Flood and river rescue operations in populated areas should be considered at the least, highly contaminated and, at the worst, potential hazardous materials scenes.*

Incident commanders should include decontamination as part of their scene structure. Personnel should consider lifejackets and wetsuits used in polluted flood water as throw away items at the end of operations. All personnel should have Hepatitis inoculations and current tetanus as well. Upper respiratory infections, skin rashes, inflamed eyes, and upper G.I. problems routinely affect rescue personnel involved in in-water rescue operations in urban areas.

13. *Don't forget "rehab"*

Protracted operations in cold water, cold weather, or rain quickly sap the efficiency of field personnel. Injuries are often the result of exposure and exhaustion - the same way they can be for firefighters exposed to heat and smoke for too long.

14. *Make sure there is plenty of water.*

While this sounds redundant, the point is that the combination of cold water and exertion will not only exhaust rescuers, but dehydrate them as well. Drinking water during flood operations can quickly become a concern. Virtually every apparatus should have food and water placed on board during such emergencies.

From the Standpoint of the emergency operations center coordinator

15. *Train and utilize multiple public information offices.*

Work to effectively use local broadcasting resources. Getting the correct information to broadcasters, with accompanying efforts to encourage information distribution while discouraging sensationalism, can aid overwhelmed public information efforts. Local broadcasting, particularly in the Los Angeles and Sacramento areas, diminished incoming calls to emergency operations centers, by advising on road closures, locations to pick up sand bags and sand, emergency shelters, areas with power losses, and other critical information.

16. *Keep dispatchers in the information "loop."*

Due to call loads, some dispatch centers around the state reported they were not receiving timely information from managers, to pass on to concerned callers.

17. *Use the I.C.S. Avoid micromanagement. Communicate.*

Some counties set up area Emergency Operations Centers to maintain control of calls and assets in the area. Problems then occurred when EOCs were unable to communicate with each other, because of busy radios or lack of dedicated phone lines.

18. *Be cautious. Activate emergency operating plans early.*

In Southern California, agencies were activated as soon as abnormal rainfall levels were determined (according to a formula that compares rainfall levels and time elapsed). In some areas of Northern California, the rise in small creeks surprised responders and dispatchers, and efforts to organize were slower. In some mountain counties EOCs were activated, even though no severe problems occurred.

A sample pre-plan or standard operating procedure (SOP) is attached to this report. It was prepared by swiftwater rescue instructor Glenn Gibbons, of the Camp Pendleton Fire Department.

ADDITIONAL READING

Hubert, July and Fuqua, Vonn. "Stormwatch: Napa Valley,"
911 Magazine, May/June, 1995.

Rigg, Nancy. "Stormwatch: Southern California,"
911 Magazine, May/June, 1995.

Seidel, G; Ojeda, E; Goldsworthy, J. "Swiftwater Rescue
Response System - Team Criteria" (Proposal). Los Angeles
City Fire Department, 1995.

"Los Angeles Inter-Agency River Rescue Operational
Procedures." Los Angeles County Fire Department,
1995.

Appendix A

Los Angeles Inter-agency River Rescue Operational Procedures

LOS ANGELES INTER-AGENCY RIVER RESCUE OPERATIONAL PROCEDURES

INTRODUCTION

The Los Angeles Flood Control System is comprised of three major waterways: Los Angeles, Rio Hondo, and San Gabriel Rivers. This system is one of the largest flood control systems in the United States. The system consists of over 450 miles of flood control channels and over 2000 miles of storm drains. This system evacuates more water during a large storm than the Mississippi River, and also carries one of the greatest flood threats. Annually, there are approximately 100 channel incidents reported. Of these approximately 40 are confirmed victims in the flood control system and rescue procedures are employed. On an average, over a ten year period, there have been six drownings associated with the flood control system.

BACKGROUND

During the 1991/92 Calendar Year, the Chief Administrative Office and the Office of Emergency Management formed a joint Swiftwater Rescue Committee, which includes: Los Angeles County Fire Department, Los Angeles City Fire Department, Los Angeles County Department of Beaches and Harbors, Los Angeles County Sheriff's Department, and the Los Angeles County Department of Public Works. The Los Angeles County Fire Department will chair this committee. The purpose of this committee is to cooperatively:

* Develop and implement predeployment strategies for potential storms.

* Establish minimum safety equipment standards for Swiftwater Rescue Response Teams and First Responder Companies.

* Define minimum training levels for certification of Swiftwater Rescue Response Teams.

* Establish standard operating procedures for swift-water rescue incidents.

* Define and standardize swiftwater rescue equipment caches.

* Interface with regional counties along with entities in the Southern United States related to swiftwater rescue procedures, equipment, training, and tactics.

TRAINING

Personnel assigned to a Swiftwater Rescue Response Team will have a minimum certification of Swiftwater Rescue Technician I. With continued in-service training proficiency would be upgraded to that of Swiftwater Rescue Technician II. The following is a total number of trained Swiftwater Rescue Technicians, by Department.

Los Angeles County Fire   - 300 Swiftwater Rescue Tech.
Los Angeles City Fire   - 37 Swiftwater Rescue Tech.
Los Angeles County Lifeguards - 40 Swiftwater Rescue Tech.
Los Angeles County Sheriff's   -   Swiftwater Rescue Tech.

SWIFTWATER RESCUE TEAMS

The minimum staffing levels on a Swiftwater Rescue Response Team is four personnel, with the preferred staffing being six personnel. To accomplish this, each four member team would be augmented with two additional personnel from either the Los Angeles County Department of Beaches and Harbors or Los Angeles County Sheriff's Department. This will increase the efficiency and overall effectiveness of the teams capabilities. An instructor will be assigned to each deployed team, Helicopter teams will consist of two person teams along with regular halitac personnel.

Each agency also has the capability to establish a team of their own personnel that can operate independently, or in in conjunction with other agencies.

VEHICLES

County Fire Department personnel will utilize reserve Paramedic Squads. City Fire Department personnel will use reserve Rescue Ambulances or emergency vans assigned to the Bureau of Support Services. The County Lifeguards will respond in vans and trucks.

COMMUNICATIONS

Fire Department Communications will be used and a tactical frequency assigned to each incident. Lifeguards and Sheriffs augmenting each team will receive radios from the Fire Department.

PREDEPLOYMENT

Swiftwater Rescue Response Teams need to be predeployed upon storm warning predictions of:

Unsaturated Conditions - 1" rain in a 24 hour period, with an 80% prediction.
Saturated Conditions - 1/2" rain in a 24 hour period, with an 80% prediction;

Other considerations for predeployment include, controlled releases of water by the Flood Control District.

Notification of predicted storm warnings will come from the Weather Bureau to Los Angeles County Fire Department, who will alert all concerned agencies of the need to predeploy. With current weather prediction information available to the concerned Department's, approximately 12-24 hours lead time would be available for staffing and predeployment of resources. Each agency will establish internal policies for notification of personnel to be predeployed.

PREDEPLOYMENT LOCATIONS

Locations that need to be considered for predeployment include:

Los Angeles County - Fire Station 1, 23, 31, 70, 87, 107, 151, and Air Operations

Los Angeles City - Fire Station 44, 62, 86, and 88.

INCIDENT COMMAND

Command of an incident will be given to the agency having jurisdiction over the rescue. Utilization of the Incident Command System will provide a synergistic approach to mitigating the incident. A swiftwater rescue dispatch will include first responder companies, helicopters, paramedics, and swiftwater rescue response teams. The Incident Commander will designate the Command Post and Staging areas in accordance with any preplans for the affected body of water.

PREWATER PLANNING

Prewater Plans will be conducted on all waterways to determine appropriate rescue locations, channel access, hazards and physical features, type of waterway, confluences, helispot locations, staging areas, and possible control devices. This preplanning will also allow companies to establish possible anchor points for victim rescue.

EQUIPMENT

Equipment will consist of personal and team equipment.
Personal equipment will include: wet suit, booties, gloves,
personal flotation device, helmet, whistle, knife, strobe
light, flashlight, and equipment bag. The team equipment
will include: swim fins, rescue board, throw bags, technical
rescue ropes, technical rescue hardware, line gun, hose
inflator, cyalume light sticks, first aid kits, flotation
rings, inflatable rescue boat, and a rescue curtain.

STANDARD OPERATING PROCEDURES

Once personnel are at there assigned locations, inventory
checks will be conducted. The apparatus is to be checked,
equipment made ready for a response, radio and
communications procedures made operational, and notification
made to the Dispatch Center of the Team's availability to
respond. The Dispatch Center will assign Team Designations
and issue tactical channels.

The initial actions of all personnel assigned to the Swift
Water Rescue Team will also include reviewing prewater
plans, assigning team positions, establishing and reviewing
safety plans, in-service team training, coordinated training
with first responder companies. Response Teams need to
drive the channels to assess hazards, velocities, volumes,
keep abreast of weather conditions and forecasts, and
respond to incidents when dispatched. These procedures will
be adhered to, until predeployment is cancellad.

EMERGENCY OPERATIONS

The strategic priorities are protection of life, rescuer
safety, and effect rescues. Notifications will be made to
any Department that may be involved in the incident. Other
priorities that will be handled include assistance to
survivors, agency liaison, public information, and scene
control.

Arrival on Scene - responsible parties or witnesses need to
be located and kept in a secure location. Appropriate
rescue information needs to be identified: do we in-fact
have a victim(s), point last scene, where victim(s) is
heading, description, mental status, etc. This will allow
for additional resources and interagency notifications. The
first company on scene needs to assess hazards and relay to
all dispatched companies. Hazards include: volume of
water, velocity, temperature, debris, low head dams,
hydraulics, and depth of water. At this time it will be
determined if we have a rescue or a recovery.

Attachment to the Los Angeles Inter-Agency River Rescue
Operational Procedures include:

    Incident Command of a Swift Water Incident
    Swiftwater Predeployment Request Worksheet
    Swiftwater Equipment check list
    Swiftwater Team Position check list

Los Angeles County Inter-Agency Swiftwater Rescue Committee

A:SW-OPS.DOC

Pre-Rescue Operations - the general area surrounding the incident needs to be made safe. This includes: moving all civilians and rescue personnel not properly equipped to a safe area. Upstream Safety shall be positioned at all incidents to ensure rescue personnel are warned of dangerous debris in the water that could impact the rescue. Downstream Safety shall be established to provide protection for both rescuers and victims, Any rescue personnel working on a river bank or in a position where they could come in contact with the water shall wear a personal flotation device and a lightweight helmet. Under no circumstance shall rescuers wear turnouts and fixed brim helmets near the incident. Rescuers are not to tie ropes around any part of there body, when they are in a position to come in contact with the water. Under no circumstance are lives to be placed at risk for a body recovery, sandbagging operation, or any other non-life saving operation. Low head dam rescues are to be done with extreme caution.

Rescue/Recovery Operation - include the plans for removal of the victim(s). Rescue operations will be made from the lowest risk to the highest risk. This will be done in all cases. By ensuring the lowest risk is utilized will increase the safety to all rescuers on scene. It is recognized that low risk operations are not always possible, but they need to be considered first. The rescue options selected will be communicatad to ail concerned to ensure a safe operation. The order for Swift-water Rescue Methods are:

> TALK - the victim(s) into self-rescue
> PROVIDE FLOTATION - give the victim(s) some means of
>   flotation
> REACH - pike pole, inflated hose lines, ladder, tension
>   diagonal, rescue curtain, shepards hook
> THROW - throw bags
> ROW - tethered boat systems, tyrolean, etc.
> GO & TOW - swimming rescues
> HELO - utilized anywhere in the above operations

The adherence to the above procedures will assist all rescue personnel in a safe, effective, and efficient rescue/recovery. The overall goal of this operations procedure is to provide additional safety and rescue equipment to areas most likely to be affected by swiftwater incidents. By augmenting first responder companies with technical Swiftwater Rescue Response Teams we can provide that advanced level of rescue and safety needed in mitigating river rescue incidents.

Appendix B

Swiftwater Incident Command System

SWIFTWATER INCIDENT COMMAND SYSTEM

INCIDENT COMMANDER

SAFETY OFFICER
INFORMATION OFFICER
LIAISON

PLANNING

RESTAT/SITSTAT
TECHNICAL INFORMATION

OPERATIONS

LOGISTICS

GROUND SUPPORT
COMMUNICATIONS

AIR OPERATIONS

COMMAND
SEARCH
RESCUE
MEDICAL

MEDICAL GROUP

TRIAGE
TREATMENT
TRANSPORTATION

SWIFTWATER GROUP

SEARCH TEAM

PRIMARY
DETAILED

RIVER RESCUE TEAM

LEADER
RESCUER
TECHNICAL
SPECIALIST(S)
UPSTREAM SAFETY
DOWNSTREAM SAFETY

FIRST RESPONDER UNITS

FLOTATION
LAND BASED RESCUE
TECHNICAL SUPPORT
SEARCH

ICS-SWIFTWATER
GES

Appendix C

"Camp Pendleton Fire Department"
"Swiftwater Rescue"
Standard Operating Procedures

"CAMP PENDLETON FIRE DEPARTMENT"

"SWIFT WATER RESCUE"

STANDARD OPERATING PROCEDURES

"CAMP PENDLETON FIRE DEPARTMENT"

SWIFT WATER RESCUE

STANDARD OPERATING PROCEDURES

1.0 PURPOSE

1.1 The purpose of this S.O.P. is to provide personnel with a basic standard plan for handling swiftwater rescue incidents. Emphasis is placed on safety of rescuers prior to committing to an attempted rescue.

2.0 SCOPE

2.1 This S.O.P. covers the basic steps or milestones to be achieved at all swiftwater rescues.

3.0 DEFINITIONS

3.1 The Swiftwater Rescue Program Coordinator will be the Division Chief assigned.

3.2 SWIFTWATER RESCUE refers to the rescue of live victims from flowing water in which the current, depth, floating or submerged hazards, contamination, or other risks exist for rescuers or victims. This S.O.P. does not refer to rescues from static water sources such as lakes or swimming pools.

3.3 RESCUE refers to the acts involved in removing a live person from a place of danger to a place of safety.

3.4 RECOVERY refers to the retrieval of a deceased person. Rescuers must assess conditions to determine if they are performing a RESCUE or a RECOVERY. Rescuers must not endanger themselves or others in the performance of a recovery. Recovery should be coordinated with representatives of PMO and/or Coroner's office, Coast Guard, Lifeguards, etc.

4.0 SWIFTWATER RESCUE STANDARD PROCEDURES

4.1 This S.O.P. is not a substitute for swiftwater rescue training. It is intended to provide general guidelines for swiftwater rescue response and to emphasize safety issues related to these calls.

4.2   As in all responses, priority for rescue must be:

      1.   **SELF RESCUE**
      2.   **RESCUE OF FELLOW RESPONDERS**
      3.   **RESCUE OF VICTIM(S)**

5.0   SWIFTWATER RESCUE TEAM LEVELS:

5.1   LEVEL "A" TEAMS

5.2   When activated, a Level "A" team provides a minimum of five and a maximum of ten specially trained firefighter personnel for immediate response to swiftwater rescue emergencies.

5.3   The level "A" team leader shall be a Fire Captain or Rescue Officer listed for the SRT Team Leader position and shall be at least SRT-2 certified.

5.4   The program coordinator shall select team members bases on need and availability.

5.5   The level "A" team shall consist of a minimum of five personnel including the Team Leader. Each of these five personnel shall be currently certified to the level of Swiftwater Rescue Technician 1, EMT and CPR.

5.6   At the discretion of the Program Coordinator, this five person team may be augmented by five additional personnel, each of whom must be currently certified to the level of SRT-1, EMT and CPR.

5.7   Level "A" teams may be activated and staged as needed to meet local rescue problems during periods of heavy precipitation.

5.8   A maximum of twenty-five (25) personnel will be listed in level "A" team membership. To remain listed, each member must re-certify their SRT-1, EMT and CPR every two years.

5.9   LEVEL "B" TEAMS

5.10   A level "B" team provides a minimum of three trained and equipped personnel to assist in swiftwater rescue problems. Trucks and selected engines will be equipped and trained to this level. Level "B" teams may be required to assist level "A" teams.

5.11   The level "B" team shall consist of all or part of selected engine companies and shall have an understanding of rigging and other setups required by level "A" teams.

5.12 Level "B" teams may start a swiftwater rescue but in no case attempt to perform at a level for which they have not been trained. They may perform LOW RISK swiftwater rescues. Such rescues would include the Talk, Reach, and Throw levels of rescue, and low risk wade-out rescues where the hazards of foot entrapment or being swept away are determined to be minimal. Level "B" teams should not do swimming rescues. Engine company personnel will be equipped and trained to the Swiftwater Rescue First Responder level.

5.13 In order to provide the greatest chance for successful rescues, it is possible to have several units respond to a single rescue incident. This may include the closest engine, closest level "B" team, or closest level "A" team.

5.14 The incident command system shall be utilized in all cases, with the IC and/or Operations Chief making the determination of which resources to utilize after completion of an incident size-up.

5.15 In making the decision about which resources to utilize the IC must consider upstream safety, downstream back-ups and alternative rescue plans. When in doubt, it would be best to continue incoming resources to the scene, stage them at the scene if there is not an immediate use for them, and release them as soon as the victim(s) have been rescued.

6.0 LEVEL OF RESCUE RISK

6.1 Swiftwater rescue activities can be categorized in order of risk to the rescuer as follows: TALK, REACH, THROW, ROW, GO, TOW, HELO.

6.2 Rescuers should evaluate the potential of each activity before proceeding to the next, riskier activity.

6.3 TALK: If the victim is capable of safely rescuing himself and merely needs instruction from rescuers, this may be done.

6.4 REACH: A hand, pike pole, or other item can be reached to the victim by a rescuer on shore.

6.5 THROW: A throwbag or rope can be thrown to the victim who is then pulled to safety by the rescuer.

6.6 ROW: If available, a boat can be used to reach the victim.

6.7 GO, TOW: Swimming to perform rescues is hazardous to both the rescuer and the victim. These rescues should be done by trained and equipped personnel only if other efforts fail.

6.8 HELO: Helicopter rescues can be very hazardous and have a high potential for disaster. This is a last option rescue, to be considered only after eliminating all other options, weighing the "Risk-vs-Save" factor, and only for rescues, never for recoveries.

6.9 Personnel lacking equipment, training, and/or adequate resources should limit themselves to the Talk, Reach, and/or Throw options. If additional help is needed, request it immediately.

6.10 Level "B" teams should restrict themselves to slow moving water "wade-out" style rescues in addition to the Talk, Reach and Throw options.

7.0 <u>15 ABSOLUTES</u> AND SAFETY RULES

7.1 Personnel should operate within their scope of training, equipment, experience, and capability.

7.2 Always wear an approved personnel flotation device (PFD), when working in the water or within ten feet of the water.

7.3 <u>Never wear turnout gear, turnout boots or wildland or structure helmets when working in the water or within ten feet of the water.</u> Each of these items makes swimming in moving water more hazardous. Given a choice between a fire helmet and no helmet, always go with no helmet.

7.4 Always deploy upstream spotters above the location of rescue operations; ideally on both sides of the river. Spotters must immediately warn others of floating or submerged debris which may pose a threat to rescuers.

7.5 <u>The priorities at the scene are always self-rescue first, the rescue and security of fellow rescuers second, and the victims last.</u>

7.6 Always have a back-up plan, and have the resources present or enroute to accomplish this plan immediately.

7.7   Always have multiple downstream backups. This may require staging companies at downstream bridges in addition to the use of throwbag and swimmer backups downstream at the rescue site.

7.8   Always keep it simple. Safety and simplicity are key elements to successful rescues. Avoid complex rope systems unless they are truly necessary.

7.9   Always use the right equipment. Inadequate or inappropriate equipment has led to a number of failed rescues.

7.10   Never put your feet down if swept away and swimming. Keep feet up and pointed downstream to protect yourself from foot entrapments and from obstructions in the river. Because of the dangers inherent in swimming in swiftwater only personnel properly trained and equipped should do so.

7.11   Never count on the victim to assist in his own rescue. Panic, hypothermia, confusion, or injury may prevent the victim from following simple commands. Be prepared to perform the rescue without assistance from the victim.

7.12   Never tie a rope around a rescuer. The rescuer must be able to swim free of the rope if the situation changes. The only exception is the use of a blow-out ring or a live bait ring on a personal flotation device by a swimmer trained in this evolution. (This is a high risk option to be avoided if possible).

7.13   When tensioning a line across the river to be used for in-water safety or walking stream crossings, never tension the line at right angles to the current. Furthermore, never clip into such a line directly. (This rule does not apply to high line systems for boat handling or litter evacuation). Run the line at a 45 degree angle to the current vector so the current will assist rather than hinder your travel. Hold the rope or hold a carabiner and strap clipped to the rope, but do not tie yourself into the rope.

7.14   While tensioning a line across the river, never stand inside the bight, and always stand on the upstream side of the rope. Should the system fail, this rule reduces the likelihood of injuries to the rescuer.

7.15   Always be pro-active. Plan ahead with pre-plans, training, and drills. Order resources as soon as they appear to be necessary, they can always be staged or canceled.

8.0   SEARCHES

8.1   In any case, where the victim(s) have not been immediately located, the IC should:

1.   Perform a size-up; Determine number, age, description, condition and last seen location of victims and their vehicles.

2.   Establish ICP and keep witnesses there.

3.   Isolate the river and deny entry to untrained and/or unequipped personnel and civilians. Do not allow additional victims to be created.

4.   Request additional resources. This may include a Chief Officer, Level "A" and/or "B" teams, Sheriffs Department for search teams, search dogs, helicopters, SCUBA team (If in the recovery mode), chaplain (to assist victim's family), PMO, Medics, TIPS, etc.

5.   Request ECC to dispatch units to downstream river access points where floating victims might be rescued. Consider lead time, response time, and the speed of the current to determine how many points and how far downstream to cover.

6.   Perform a Hasty Team downstream search, but keep rescuer safety as a priority and don't over commit your resources.

7.   Prepare ICS 201 and brief incoming resources. Keep track of actions taken and times if possible.

8.   After searching an automobile or structure, indicate the search by painting an "OK" on the vehicle or structure, and paint the date and time of the search and who performed the search (if possible).

9.0   HELICOPTERS

9.1   Helicopters present a high risk option to both rescuers and victims. Any use of helicopters must be preceded by addressing safety issues. Even experienced helicopter crews may be unfamiliar with the effects of river currents on their ship and crew. Before a helicopter rescue is attempted with an unknown pilot and aircraft, the IC must balance the life threat against the pilot and crew expertise and aircraft capabilities. Coordinate your planned use of the helicopter with the pilot before committing to the plan. Use helicopters only as a last resort.

9.2   Helicopters can be used for (in order of increasing hazard):

1.   Area reconnaissance to locate victims, hazards, river hydrology features, crossing points, etc.

2.   Scene lighting at night.

3.   Transporting rescuers and equipment from one helispot to another.

4.   Transporting victims rescued by ground teams.

5.   Deployment of a rope across a river from two points on shore.

6.   Rappels.

7.   Picking up a victim from a stable location in the river (roof of a car, island, etc). Note that as the speed of the current increases, so does the risk to all personnel involved.

8.   Picking up a victim floating in the water. (This highly hazardous rescue should be avoided if safer options exist).

9.3   In all helicopter operations, safety and effectiveness will be enhanced if an experienced and trained swiftwater rescue technician is riding on the helicopter and has good communication with both the pilot and Incident Commander. The ability to accurately interpret river hydrology features and to communicate the significance of these make this spotter a valuable safety asset.

10.0   MEDICAL CONSIDERATIONS

10.1   Dehydration, hypothermia, near-drowning syndrome, drowning, and all manner of trauma are common water related injuries. In addition, contaminated water may contribute to infections. Be prepared, watch for symptoms, and provide care as needed.

10.2   If a victim or rescuer becomes disoriented or is unable to drink fluids, this patient must be transported to a medical facility for evaluation. Minor mental impairment may be a symptom of an injury which will progressively worsen and may become fatal. (Near drowning "parking lot syndrome").

10.3   Even in cool weather, all personnel must keep fluid intake high to battle the effects of heat, cold, and dehydration.

11.0 PREPLANS

11.1   River and Flood Channels should be preplanned by local units to determine:

      1.   Travel routes and all access points to both sides (note locked gates and dirt sections where mud will impede progress).

      2.   Hazards such as low head dams, strainers, etc.

      3.   Underground sections.

      4.   River crossing points.

      5   Helispots and aerial hazards (power lines, etc).

      6.   Current speed and vector.

      7.   Eddies and pools (good rescue points).

      8.   Downstream companies and staging locations.

      9.   Other noteworthy features.

12.0 INCIDENT DOCUMENTATION

12.1   A CPFD form - River Rescue Incident Report Form must be completed on each victim rescued, whether or not they are transported to a facility. A copy of this form is attached and listed as appendix A.

12.2   A brief narrative of the type of rescue performed and the names of the rescuers should be included on the report form.

12.3   In the event of a fatality, the coroner may wish to inspect any technical rope rescue system that was used in the rescue effort. The IC should coordinate this with the coroner, or at least document the type and location of the system used and the results.

13.0 ALERT STATUS

13.1   There will be four (4) Alert Conditions. These conditions will be consistent with the Base Flood Plan.

13.2   Condition Four (4):
Condition-4 will exist when weather conditions are severe enough and water levels are swift or high enough or have the potential of being swift or high enough so that a potential swift water rescue may occur.

13.3   When the fire department enters into condition-4, a Chief Officer will advise the Emergency Communications Center (ECC), whereas, the Dispatcher on duty will immediately notify the on duty SWR Team Leaders and Assistant Team Leaders.

The Team Leaders will then put together a five (5) member team for each Division, available for immediate response. If there are not enough members on duty to support a five member team for each Division, then all available team members on duty will be placed on alert to respond. The team leaders will notify each team member's immediate supervisor and the member of their status. The team leaders will also notify the Division 1 & 2 Chiefs and the Dispatcher. The intent of Condition-4, is only to identify which members will be part of the initial response teams and to let the appropriate people know who they are.

13.4   Conditions (3), (2) & (1):
Conditions 3,2 & 1, will exist when the Base Flood Plan is in Condition 3,2 or 1. The duties and responsibilities of the Swiftwater Rescue Teams will be consistent with the Base Flood Plan. These conditions will be passed to the ECC Dispatcher. The Dispatchers will ensure that all Chief Officers and Swiftwater Rescue Team Leaders and Assistant Team Leaders are notified.

"GOOD   LUCK"
&
"SAFE   RESCUING"

Appendix D

River Rescue Incident Report Form

River Rescue Incident Report Form

Person Completing Report_____Date_____Time_____

Incident Type: Rescue____Medical____ Practice____ Assist other group___

Approximate Air Temp_____ Approximate Water Temp_____

Weather Conditions_____

Incident Location:_____

_____

Company or Personal Report Number_____

If Applicable, Reporting Party's Name_____

Address_____City_____State_____Zip_____

Home Phone(_____)_____ Work Phone(___)_____

Time incident Occurred_____Rescue Initiated_____Rescue Secured _____

Personnel at Scene: **Names   Address   Time In   Out**

_____

_____

_____

_____

_____

Incident Leader_____

Medic_____

Primary Rescuers_____

"Gofer" (Recorder, Personnel, Equipment)_____

If Applicable, complete below-

Name of Victim_____Age___Sex__DOB_____

Address_____City_____State Zip ___

Employment _____ Phone_____

Activity Participating in_____Activity Time Lost_____

Is This a Reinjury of an Old Condition?_____

Were There Other Contributing Factors? If yes, What?_____

_____

Did Equipment Contribute in any way to the Accident? How?_____

_____

Has the Injured Party Signed a Release and is it Available?_____

Has the Injuried Party Participated in this Activity at this Location Before?_____

Is the Injured Party Formerly or Presently An Employee?_____

Did the Injured Party Contribute to the Accident in any Way?_____

Did the Injured Party State that He/She Contributed to the Accident in any Way?_____

Did the injured Party Refuse First Aid or Evacuation?_____

Did Another Participant Contribute to the Injury? How?_____

Were There Other People Injured in this Accident? Describe_____

Medical Type (Use as many *as* needed):   _Chest   _Spine   _Abdomen

_Extremities   _Airway   _Chest Pain   _Blood Sugar _CVA   _Drowning

_Drugs/Alcohol _Shock   _Burns   _Smoke Inhal _Seizure   _Resuscitation

_DOA   _Head, Ears, Nose, Throat   **_Unkown**

Airway:	_Clear	_Compromised	_Obstructed	
Lung Sounds:	_Clear	**_Wheezes**	_"Wet"'	_None
Pulses:	_Full	_Thready	_Irregualr	_None
Skin color:	_Normal	_Flushed	_Pale.	_Cyanotic (Blue)
Skin Condition:	_Normal	-Cool, Clammy	_Cool; Dry	_Hot, Moist _Hot, Dry
Pupils:	_Equal, Reactive to Light	_Unequal	_Sluggish	-Non-reactive

---

Vitals	Time	Blood Pressure	Pulse	Respirations	Consciousness

Time of Onset _____ Call Rec'd _____ Enroute_____ At Scene_____ From Scene_____

Victim turned over to Ambulance_____ Victim at Hospital _____

Chief Complaint_____

Treatment Prior to Arrival_____

Brief medical history, medicines, allergies, last time saw doctor, problems with heart, lungs, diabetes, epilepsy, 'asthma, emphysema, bronchitis?_____

Narrative-if medical include treatment, equipment used, physical survey. If rescue, include times, personnel, equipment, map if necessary. Photos?

_____
_____
_____
_____
_____
_____
_____
_____
_____
_____
_____
_____
_____

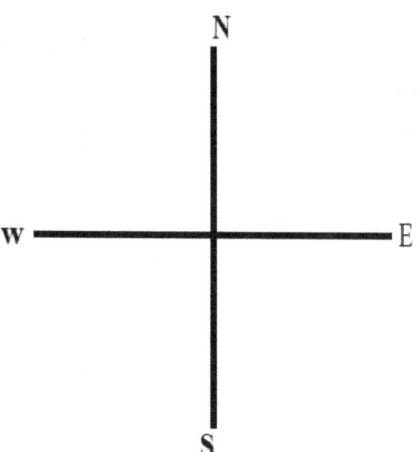

Attach supplemental narrative if necessary.

Signature   Time   Date   Place

_____